Math Grid Games

WRITTEN BY
DAVID THOMAS

World Teachers Press

Published with the permission of R.I.C. Publications Pty. Ltd.

First published by R.I.C. Publications Pty. Ltd., Perth, Western Australia.

Printed in the United States of America.

Order Number 2-5064
ISBN 1-885111-78-9

A B C D E F 98 99 00 01

395 Main Street
Rowley, MA 01969

Foreword

Mathematical board games can be used to great effect in the classroom and their motivational value has long been recognized by teachers. In order to avoid many of the problems associated with complex games including lost playing pieces, unusual formats and complicated rules, *Math Grid Games* have been designed with a unique concept in mind. They use:

a) only dice and counters as playing equipment;
b) simple, consistent grid formats; and
c) straightforward rules, printed alongside each game.

Math Grid Games can be photocopied and are written to support modern mathematics curricula. Each game can be adapted for a wide range of teaching strategies, from small group work to whole-class activity.

The games include concepts from the three major areas of mathematics - number, space and measurement.

Math Grid Games can provide practice, reinforcement and even the introduction of specific concepts in a stimulating way.

Teacher Information

Classroom Applications for Math Grid Games

Concept Development
With the practical nature of these games, it is possible for the whole class to play one game by duplicating sufficient sheets. This allows the game to become an instructional tool used as an integral part of the class math program.

Extension/Remediation
The games can be used to extend or remediate, depending on the level each student is at. Because game instructions and rules are simple, these games can be used in small group situations with students of similar ability to reinforce teaching points and also to extend more able students.

Learning Centers
The games can be used as an excellent math learning center. By duplicating, coloring and then laminating each game, a set of 48 games is available for constant use throughout the day. As there are very few materials required and the games are short in duration, this requires minimal class management.

Enlargement
To enhance the appeal of the games and to assist less able students, the games can be enlarged on a photocopier to provide a larger playing area.

Spotty

You Need

2 dice
1 counter for each player

Rules

Take turns to roll one die. Move your counter. If you land on a white spot, throw both dice and add them together.
Move forward this number of squares and roll again.
If you land on a black spot, throw both dice and subtract the smallest number from the largest number (this is called finding the difference).
Move forward this number of squares and roll again. If you land on a blank square, stay on that square until your next turn.
The first player to reach 36 is the winner.

When You Have Finished

Talk about these questions:
What is the biggest difference you could have made in this game?
What is the greatest number of spaces you could move in one turn in this game?

36	35	34	33	32	31
25	26	27	28	29	30
24	23	22	21	20	19
13	14	15	16	17	18
12	11	10	9	8	7
1	2	3	4	5	6

Big or Small

You Need

1 die
1 counter for each player

Rules

Take turns to roll the die. Move your counter. If you land on a circle, look at the number inside it. If the number in the circle is bigger than the number of the square it is in, roll the die again.
If the number in the circle is smaller than the number of the square it is in, your turn is finished. If you land on a blank square, stay on that square until your next turn. The first player to reach 36 is the winner.

When You Have Finished

Look at the number 36 on the last square.
Is this the biggest number on the board?

What s That Number?

You Need

1 die
1 counter for each player

Rules

Take turns to roll the die. Move your counter. If you land on a square with a 'missing number problem', find out the missing number and move forward that number of squares.
When you land on a blank square that is the end of your turn.
The winner is the player to reach 36 first.

When You Have Finished

Which numbers could fit this problem?

□ + □ = 8

36	35	34 2 + 3 = □	33	32 4 + □ = 5	31
25	26 2, 4, 6, □	27	28 □ + 2 = 5	29	30 7 - □ = 3
24 4 + □ = 6	23	22 7 - 2 = □	21	20 3 + □ = 5	19
13	14 1, 2, □, 4, 5	15	16 2, 3, □, 5, 6	17	18 2, 4, □, 6, 8
12 □ - 2 = 1	11	10 6 - 2 = □	9	8 3 - □ = 1	7
1	2 6 + □ = 10	3	4 4 + 3 = □	5	6

Using A Calculator

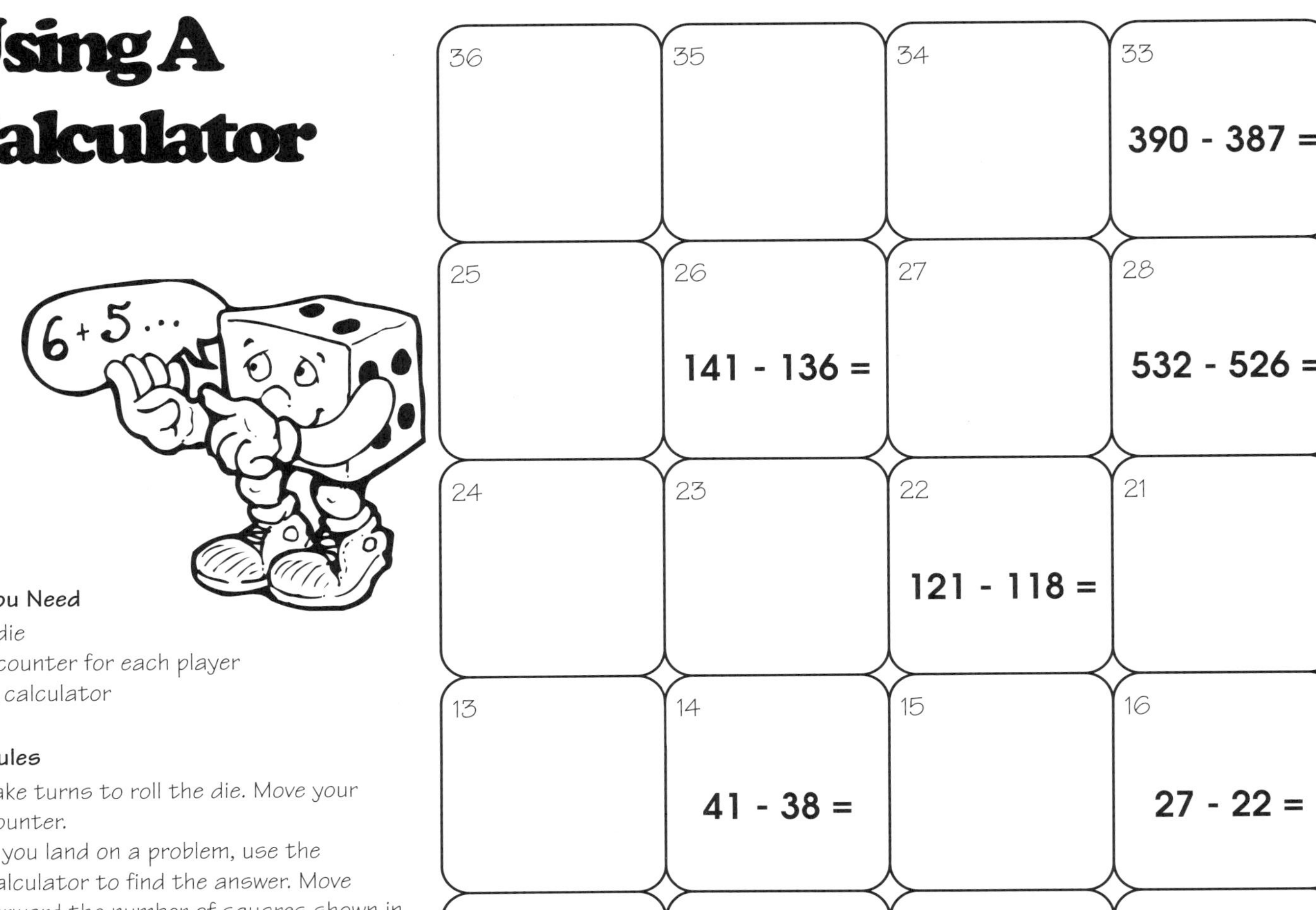

You Need

1 die
1 counter for each player
A calculator

Rules

Take turns to roll the die. Move your counter.
If you land on a problem, use the calculator to find the answer. Move forward the number of squares shown in your answer.
End your turn there. If you land on a blank square, stay on that square until your next turn.
The first player to reach 36 is the winner.

When You Have Finished

If you add the numbers instead of taking them away, what would be the largest number you could make?
Use the calculator to check your answers.

36	35	34	33 390 - 387 =	32 221 - 219 =	31
25	26 141 - 136 =	27	28 532 - 526 =	29	30 318 - 317 =
24	23	22 121 - 118 =	21	20 101 - 96 =	19
13	14 41 - 38 =	15	16 27 - 22 =	17	18 52 - 49 =
12 81 - 76 =	11	10 15 - 10 =	9	8 45 - 44 =	7
1	2 56 - 55 =	3	4 39 - 35 =	5	6 81 - 78 =

Halves and Quarters

You Need

1 die for each player
1 counter for each player

Rules

Take turns to roll the die and move your counter.

Some squares have a fraction like this on them. $\frac{1}{2}$
Others have a shape which shows a fraction shaded like this.
If you land on one of these squares you must say whether the fraction is a half or a quarter. If it is a half, go forward TWO squares. If it is a quarter, go forward FOUR squares.
If you land on a blank square stay on that square until your next turn.
The first player to reach 36 is the winner.

When You Have Finished

How many quarters does it take to make a half?
How many quarters does it take to make a whole?

36	35 $\frac{1}{2}$	34	33	32	31
25	26	27	28	29	30
24	23 $\frac{1}{4}$	22	21	20	19
13 $\frac{1}{2}$	14	15	16	17 $\frac{1}{4}$	18
12	11	10	9	8	7
1	2	3	4	5 $\frac{1}{2}$	6

World Teachers Press

Short Tables 1

▶ **You Need**

1 die for each player
1 counter for each player

▶ **Rules**

Take turns to roll the die.
Move your counter forward that number of squares. If you land on a number circle, then **multiply** the number on the die by the number in the circle.
For example,
- you throw a four
- you then move four spaces
- you land on a blank space - do nothing
- you land on a numbered square (say 2) - you multiply 2 times the number on the die i.e., 2 times 4 = 8 spaces, you move 8 spaces and your turn is complete.

The first player to reach 100 is the winner.

▶ **When You Have Finished**

How many steps of 5 or 10 would it take to get from 1 to 100?

100	99	98	97	96	95	94	93	92 (2)	91
81	82 (2)	83	84 (2)	85	86 (2)	87	88 (2)	89	90 (2)
80 (3)	79	78 (2)	77	76 (3)	75	74 (2)	73	72 (3)	71
61	62 (2)	63	64 (3)	65	66 (2)	67	68 (3)	69	70 (2)
60 (3)	59	58 (2)	57	56 (3)	55	54 (2)	53	52 (3)	51
41	42 (2)	43	44 (3)	45	46 (2)	47	48 (3)	49	50 (2)
40 (3)	39	38 (2)	37	36 (3)	35	34 (2)	33	32 (3)	31
21	22 (2)	23	24 (3)	25	26 (2)	27	28 (3)	29	30 (2)
20 (3)	19	18 (2)	17	16 (3)	15	14 (2)	13	12 (3)	11
1	2 (2)	3	4 (3)	5	6 (2)	7	8 (3)	9	10 (2)

Short Tables Practice

36	35	34	33	33	33
25	26	27	28	33	30
24	23	22	21	20	33
13	14	15	16	33	18
12	11	10	9	8	33
1	2	3	4	5	6

You Need

2 dice
6 counters for each player

Rules

Take turns to roll both dice.

Multiply the two dice numbers
e.g., (4 X 3 = 12).
Place a counter on that square.
If another player has a counter on that square, take that counter off.

The winner is the first player to have all six counters placed on the board.

When You Have Finished

Why have some of the squares been shaded?

Addition or Subtraction?

You Need

1 die, 1 counter for each player

Rules

When you land on a square with a letter (A, B, C, D, E, F) you must read that question below and decide whether it is a subtraction or addition problem. If you need to subtract, then move backwards the number shown on the die. If you need to add, move forward the number on the die.

A) John has 50¢ and he spends 25¢, how much money does he have left?

B) Sally has 35¢ and her mom gives her 20¢ more, how much money does she have altogether?

C) 15 red cars and 12 blue cars drive past the school. How many cars is this altogether?

D) A boy has 18 candies and he gives his friend 6 of them. How many candies does he have left?

E) There are 14 girls and 16 boys in my class. How many children all together?

F) The teacher has 15 balls in a box and he gives out 6 balls to use at recess. How many balls are left in the box?

If you land on a blank square stay on that square until your next turn.

The winner is the first player to reach 100.

When You Have Finished

See who can work out the answers to the six questions the quickest. Write down your answers and then check them.

100	99	98	97 F	96	95 E	94	93 D	92	91
81	82	83 F	84	85 A	86	87 B	88	89 C	90
80	79 E	78	77 D	76	75 C	74	73 B	72	71
61	62	63 D	64	65 E	66	67 F	68	69 A	70
60	59 C	58	57 B	56	55 A	54	53 F	52	51
41	42	43 B	44	45 C	46	47 D	48	49 E	50
40	39 A	38	37 F	36	35 E	34	33 D	32	31
21	22	23 B	24	25 A	26	27 B	28	29 C	30
20	19 F	18	17 E	16	15 D	14	13 C	12	11
1	2	3	4	5	6	7 A	8	9 B	10

Little Money Problems

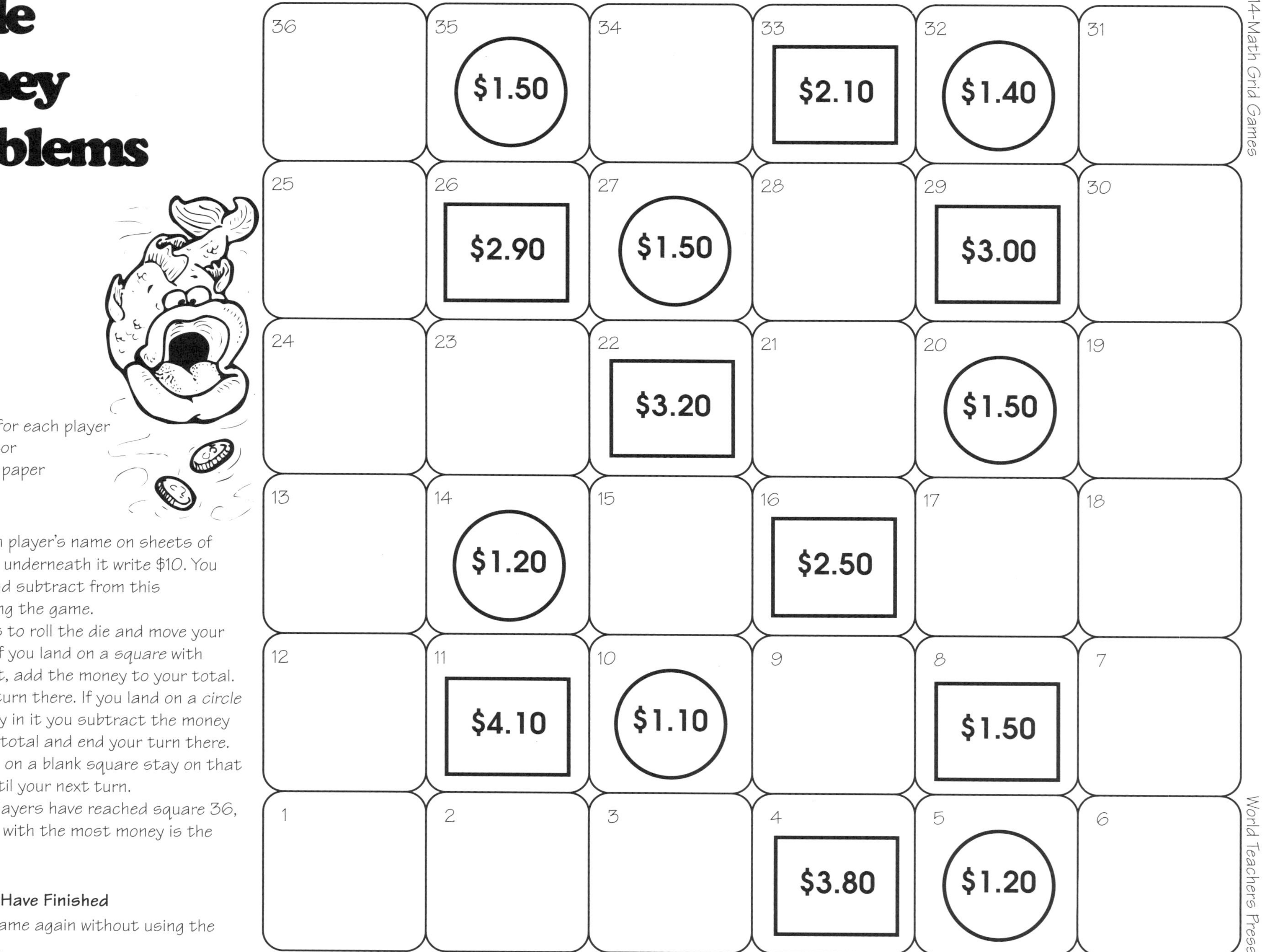

You Need

1 die
1 counter for each player
A calculator
Pencil and paper

Rules

Write each player's name on sheets of paper and underneath it write $10. You will add and subtract from this total during the game.

Take turns to roll the die and move your counter. If you land on a *square* with money in it, add the money to your total. End your turn there. If you land on a *circle* with money in it you subtract the money from your total and end your turn there.

If you land on a blank square stay on that square until your next turn.

When all players have reached square 36, the player with the most money is the winner.

When You Have Finished

Play the game again without using the calculator.

Negative Numbers

You Need

1 die
1 counter for each player
A calculator for each player

Rules

Take turns to roll the die and move your counter. If you land on a square with a problem in it use your calculator to solve the problem. Move forward the number shown in the answer if it is a positive number. Move back the number shown in the answer if it is a negative number. If you land on a blank square stay on that square until your next turn. The winner is the first player to reach 100.

When You Have Finished

Change the sums to addition and work them out with your calculator. Which is the largest answer?

100	99	98 92 - 95	97	96 84 - 83	95	94 71 - 68	93	92 68 - 63	91
81	82 53 - 48	83	84 76 - 71	85	86 84 - 81	87	88 79 - 82	89	90 81 - 84
80 52 - 49	79	78 41 - 39	77	76 59 - 62	75	74 101 - 98	73	72 98 - 101	71
61	62 92 - 95	63	64 84 - 83	65	66 71 - 68	67	68 68 - 63	69	70 72 - 69
60 53 - 48	59	58 76 - 71	57	56 84 - 81	55	54 79 - 82	53	52 81 - 84	51
41	42 52 - 49	43	44 41 - 39	45	46 101 - 98	47	48 98 - 101	49	50 81 - 84
40 92 - 95	39	38 84 - 83	37	36 71 - 68	35	34 68 - 63	33	32 72 - 69	31
21	22 53 - 48	23	24 76 - 71	25	26 84 - 81	27	28 79 - 82	29	30 81 - 84
20 41 - 45	19	18 52 - 49	17	16 38 - 35	15	14 44 - 41	13	12 30 - 31	11
1	2 28 - 27	3	4 35 - 31	5	6 24 - 21	7	8 19 - 22	9	10 22 - 19

Long Tables Practice

You Need

2 dice
10 counters for each player

Rules

Roll both dice.
Add the dice together. This is your first number. Roll the dice again and add them together. This is your second number.
Multiply your first and second number.
Place a counter on that square.
For example, rolling 4 and 2 will give you a first number of 6. Rolling 2 and 3 next will give you a second number of 5.
Multiply 6 and 5 to give you 30.
Place a counter on square 30.
If another player has a counter on that square, replace it with yours. The first player to have all ten counters on the board is the winner.

When You Have Finished

Why have some of the numbers been shaded?

100	99	98	97	96	95	94	93	92	91
81	82	83	84	85	86	87	88	89	90
80	79	78	77	76	75	74	73	72	71
61	62	63	64	65	66	67	68	69	70
60	59	58	57	56	55	54	53	52	51
41	42	43	44	45	46	47	48	49	50
40	39	38	37	36	35	34	33	32	31
21	22	23	24	25	26	27	28	29	30
20	19	18	17	16	15	14	13	12	11
1	2	3	4	5	6	7	8	9	10

Place Value

You Need
1 die
1 counter for each player

Rules
Remember - a digit is one part of a number. It can be 0, 1, 2, 3, 4, 5, 6, 7, 8, 9.

Take turns to roll the die and move your counter. If you land on a square with a number, throw the die again. If the digit on the die matches one of the digits in the number then move forward:
10 places if it is in the 1,000's column;
5 places if it is in the 100's column;
3 places if it is in the 10's column; or
1 place if it is in the 1's (units) column.
Then roll again.
If you land on a blank square stay on that square until your next turn.
The winner is the first player to reach 100.

When You Have Finished
What is the biggest number you can make if you use all the digits on a die once?

100	99	98 1,354	97	96	95 2,461	94	93	92 6,543	91
81	82 2,615	83	84	85 3,512	86	87	88 4,612	89	90
80 5,321	79	78	77 4,215	76	75	74 1,362	73	72	71 2,346
61	62	63 5,612	64	65	66 5,132	67	68	69 6,214	70
60	59 1,235	58	57	56 5,163	55	54	53 3,165	52	51
41 3,124	42	43	44 1,256	45	46	47 6,325	48	49	50 5,413
40	39	38 3,265	37	36	35 6,531	34	33	32 5,214	31
21 2,561	22	23	24 6,531	25	26	27 2,314	28	29	30 2,165
20	19	18 2,516	17	16	15 6,324	14	13	12 2,516	11
1	2 5,634	3	4	5 3,216	6	7	8 1,463	9	10

Remove the Digit

You Need
- 1 die
- A calculator
- 1 counter for each player

Rules

Take turns to roll the die and move your counter. Move forward that number of spaces. If you land on a hexagon, check and see if the number on the die is the same number as one of the digits in the hexagon. If it is, use your calculator to remove the digit or change it to zero.

For example, if you throw a 5 and land on 5,263, subtract 5,000 to leave 263. If you had thrown a 2, then subtract 200 to leave 5,063. If you do this correctly, then roll again and move on. If you land on a blank square stay on that square until your next turn.The first player to reach 100 is the winner.

When You Have Finished

What is the smallest number on the board?
What is the largest number on the board?

100	99 23,461	98	97 123	96	95 2,134	94	93 56,214	92	91 46,312
81 21,654	82	83 324,651	84	85 231,654	86	87 34,625	88	89 42,163	90
80	79 42,165	78	77 312,465	76	75 546,213	74	73 36,124	72	71 4,612
61 54,132	62	63 43,216	64	65 326	66	67 1,462	68	69 324	70
60	59 3,214	58	57 6,143	56	55 34,216	54	53 643	52	51 1,256
41 654,321	42	43 2,345	44	45 12,436	46	47 45,621	48	49 2,364	50
40	39 42	38	37 1,654	36	35 3,216	34	33 453	32	31 62
21 52,146	22	23 3,651	24	25 2,463	26	27 1,324	28	29 6,531	30
20	19 4,563	18	17 26	16	15 3,421	14	13 23,456	12	11 123,456
1 234	2	3 1,356	4	5 2,431	6	7 5,643	8	9 1,264	10

Big Money Problems

You Need

1 die
1 counter for each player
A calculator
Pencil and paper

Rules

Write down each player's name on sheets of paper and underneath that write $10. You are going to add and subtract from this figure. Take turns to roll the die. If you land on a *savings* ***square***, multiply the money by the number on the die. Add this to your total and roll again.

If you land on a *shopping* ***circle***, divide the money by the number on the die. Take this away from your total and roll again. If you land on a blank square stay on that square until your next turn.Keep playing until every player has reached 100. The one who finishes with the most money is the winner.

When You Have Finished

What is the difference between these amounts:
$3.50 $3.05 $3.5
If you saw these amounts on a calculator what would they mean?
$2.9, $4.7, $1.6, $7.4, $0.3, $.2

100	99	98 (circle) $4.80	97	96	95 (square) 15¢	94 (circle) $3.60	93	92 (square) 75¢	91
81	82 (square) 60¢	83 (circle) $2.40	84 (square) 35¢	85	86	87 (square) $1.25	88 (circle) $2.40	89	90
80	79	78	77 (square) 40¢	76 (circle) $4.80	75	74	73 (circle) $3.60	72 (square) 30¢	71
61	62 (circle) $3.60	63 (square) 95¢	64	65	66 (square) $1.10	67 (circle) $1.80	68	69	70
60	59	58 (square) 75¢	57 (circle) $4.80	56	55	54 (square) $1.50	53 (circle) $2.40	52	51
41	42 (circle) $3.60	43	44	45 (square) 25¢	46	47 (circle) $1.20	48 (square) $1.10	49	50
40 (square) 45¢	39 (circle) $4.80	38	37	36 (circle) $2.40	35 (square) 35¢	34	33	32 (circle) $3.60	31
21 (square) 65¢	22	23 (circle) $1.20	24 (square) $1.15	25	26 (circle) $4.80	27	28 (square) 35¢	29	30 (square) 90¢
20	19	18 (square) $1.20	17 (circle) $1.80	16	15 (square) 75¢	14	13	12 (square) 90¢	11
1	2	3	4	5 (square) $1.10	6	7 (square) $1.30	8	9	10

Spotty Squares

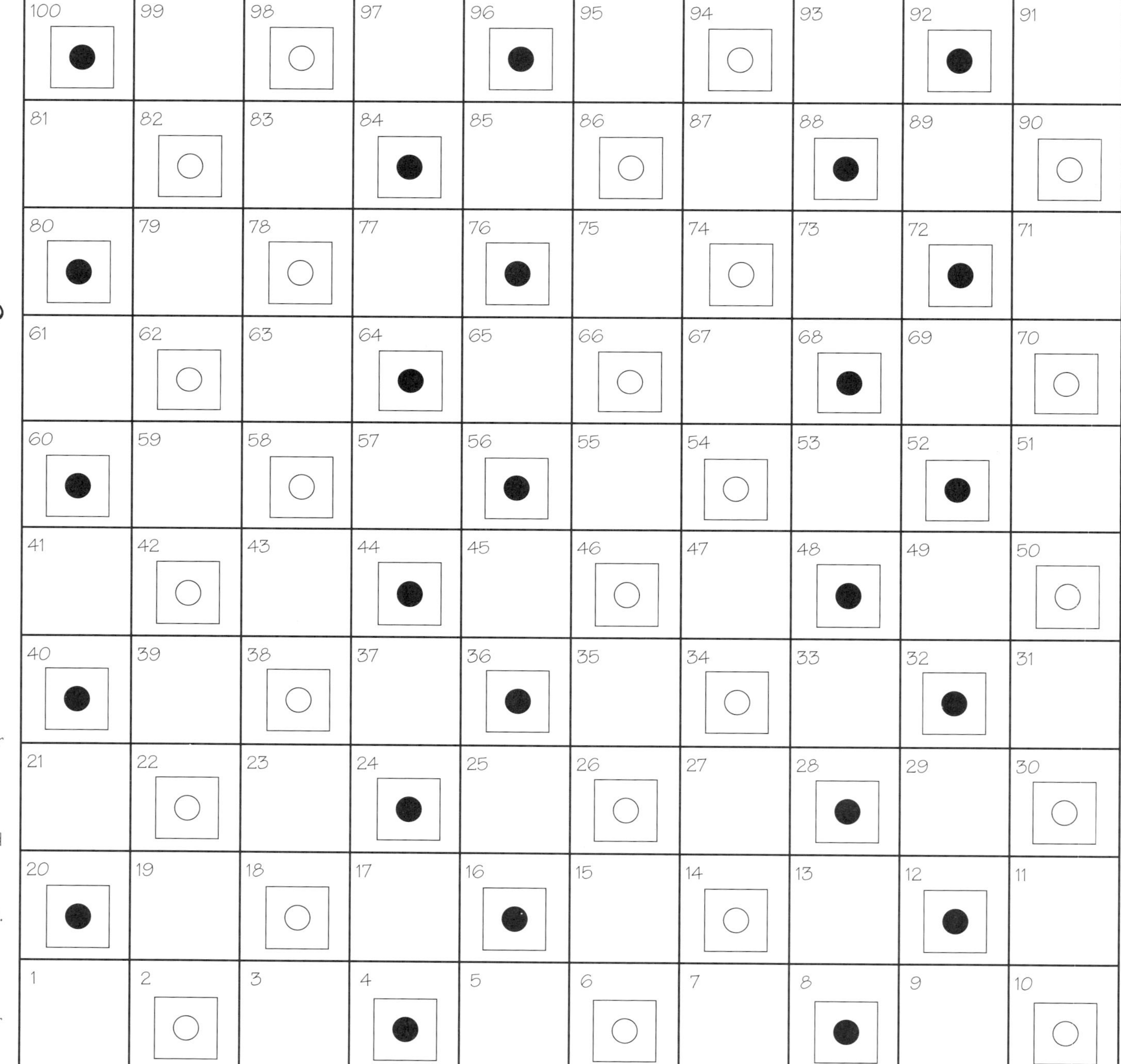

You Need

2 dice

1 counter for each player

Rules

Take turns to roll the dice and move your counter. If you land on a square with a **black spot,** throw two dice and add them together. Move forward this number and stop there even if you are on another spot.

If you land on a square with a **white spot,** throw two dice and add them together - this is your first number. Throw two dice again and add them together. This is your second number. Subtract the smaller number from the larger number (find the difference) and move forward this number. Stop there until your next turn. If you land on a blank square stay on that square until your next turn.

The winner is the first player to reach 100.

When You Have Finished

What is the largest total you can make with three dice? What is the least number of moves to get from square 1 to 100?

Mental Arithmetic

▶ **You Need**

3 dice
1 counter each
A calculator

▶ **Rules**

Throw one die and move your counter.
If you land on an *add* square or a *subtract* square then:
Throw two dice to make a number, the first die is the tens and the second is the units, (say 4 and 2 = 42); then throw them again to get the next number (say, 5 and 3 = 53).
If you are on an *add* square, find the total (42 + 53 = 95).
If you are on a *subtract* square, find the difference (53 - 42 = 11).
Work out the answer mentally and get your partner to check the answer with the calculator. If you are correct, take another turn. If you are incorrect that is the end of your turn. If you land on a blank square stay on that square until your next turn.
The winner is the first player to reach 100.

▶ **When You Have Finished**

How could this game be adapted to include multiplication and division?

100	99 −	98	97 +	96 −	95	94 +	93	92	91 +
81 +	82	83 −	84	85	86 +	87	88 −	89	90
80	79 −	78	77 +	76 −	75	74 +	73	72	71 +
61 +	62	63 −	64	65	66 +	67	68 −	69	70
60	59 −	58	57 +	56	55	54	53	52	51
41 +	42	43 −	44	45	46 +	47	48 −	49	50 +
40	39 −	38	37 +	36	35	34 +	33	32	31 +
21 +	22	23 −	24	25 −	26 +	27	28 −	29	30
20	19	18 −	17 +	16	15	14 +	13	12	11 +
1	2	3 +	4	5 −	6 +	7	8 −	9	10

From now on throw three dice each turn to get three-digit numbers

Mind Your Numbers

You Need

1 die
1 counter for each player
A calculator

Rules

Roll the die and move your counter.
If you land on a square with a number, roll the die that number of times.
Add the total of the numbers mentally as you go along. If you are correct, take another turn.
If the other player thinks you are wrong, check the answer with a calculator. If you are wrong go back 5 spaces and end your turn. If you land on a blank square stay on that square until your next turn. The winner id the first player to reach 100.

When You Have Finished

What are the largest and smallest numbers you can make in this game?
How many times can you roll the die and add without making a mistake or forgetting? Practice with your partner.

100	99	98 9	97	96	95 7	94	93	92 8	91
81	82 9	83	84	85 7	86	87	88 8	89	90
80 5	79	78	77 6	76	75	74 7	73	72	71 8
61	62	63 7	64	65	66 3	67	68	69 4	70
60	59 3	58	57	56 4	55	54	53 5	52	51 6
41 4	42	43	44 5	45	46	47 6	48	49 7	50
40	39	38 6	37	36	35 7	34	33 3	32	31
21	22 7	23	24	25 3	26	27 4	28	29	30 5
20 3	19	18	17 4	16	15 5	14	13	12 6	11
1	2	3 3	4	5 4	6	7	8 5	9	10 6

Familiar Fractions

You Need

1 die

Note: For this game you need to cover the middle dot of the number five on the die to make it a second four and change the 1 to a 12

1 counter for each player

Rules

Take turns to roll the die. Move your counter. If you land on a square with a number, make the number on your die the bottom half (denominator) of a fraction with the number one on the top (numerator). Then find this fraction of the number on the square.

For example, if you roll a 2 then your fraction will be $^1/_2$. If you have landed on a square showing the number 12, then you find $^1/_2$ of 12 which is 6. Move forward 6 squares.

If you land on a blank square stay on that square until your next turn.

The first player to reach 100 is the winner.

When You Have Finished

What are the largest and smallest numbers you could get in this game?

Put these fractions in order of size:

$^1/_2$ $^1/_6$ $^1/_4$ $^1/_3$

100	99	98	97 12	96	95	94 24	93	92	91 36
81	82	83 24	84	85	86 36	87	88	89 12	90
80	79 36	78	77	76 12	75	74	73 24	72	71
61 12	62	63	64 24	65	66	67 36	68	69	70 12
60	59	58 36	57	56	55 12	54	53	52 24	51
41	42 12	43	44	45 24	46	47	48 36	49	50
40 24	39	38	37 36	36	35	34 12	33	32	31 24
21	22	23 12	24	25	26 24	27	28	29 36	30
20	19 24	18	17	16 36	15	14	13 12	12	11
1	2	3	4 12	5	6	7 24	8	9	10 36

Inverses

You Need

1 die

1 counter for each player

Rules

Take turns to roll the die and move your counter. When you land on a square with a problem, solve it by calculating the missing number. Move forward the number given in your answer. If you land on a blank square stay on that square until your next turn.

The first player to reach 100 is the winner.

When You Have Finished

What pattern or rule have you found? Can you find numbers which fit these letters and solve the problems?

(i) y X z = p and p ÷ y = z

(ii) s + n = b and b - n = s

100	99	98 87÷29=3 29x?=87	97 76÷38=2 ?x38=76	96	95	94	93 88÷2=44 ?x44=88	92 92÷4=23 23x?=92	91
81	82 88÷4=22 22x?=88	83 54÷6=9 9x?=54	84	85 74÷2=37 37x?=74	86	87 84÷3=28 28x?=84	88	89	90 64÷16=4 16x?=64
80 78÷13=6 ?x13=78	79 90÷18=5 ?x18=90	78	77	76 75÷5=15 15x?=75	75 63÷3=21 21x?=63	74	73 84÷4 =21 ?x21=84	72	71
61	62	63 49÷7=7 7x?=49	64	65 120÷20=6 20x?=120	66	67	68 15x4=60 60÷15=?	69 12+8=96 96÷12=?	70
60	59 56÷8=7 8x?=56	58 27+9=3 9x?=27	57	56	55	54 64÷8=8 8x?=64	53 18+9=2 ?x9=18	52	51 5x6=30 30÷5 =?
41 81÷9=9 9x?=81	42	43 9x7=63 63÷?=9	44	45 100÷20=5 20x?=100	46	47 20x5=100 100÷20=?	48	49 77÷11=7 ?x11=77	50
40	39 21÷7=3 7x?=21	38	37 11-10=1 ?+10=11	36	35	34	33 18-16= 2 16+?=18	32 14÷7=2 ?x7=14	31
21 16-9=7 ?+9=16	22	23 15÷5=3 3x?=15	24	25	26 16-12=4 ?+12=16	27 7x8=56 56÷?=8	28	29	30 6x9=54 54÷?=9
20	19 17+ 5=22 22- ?=17	18	17 7 x 8 =56 56÷ ?= 7	16	15 6x?=18 18÷3=6	14	13	12 4x2=8 8÷2=?	11
1	2	3 3+2=5 5-2=?	4	5 4-3=1 1+3=?	6	7	8 24-6=18 18+?=24	9	10 15+3=18 18-?=15

Negative Thinking

So if I subtract this from that, I should get...

You Need

1 die
1 counter for each player
A calculator

Rules

Take turns to roll the die and move your counter. When you land on a circle, subtract the number in the circle from the number on your die. Don't use the calculator.
Move forward that number if your answer is a positive number.
Move back that number if your answer is a negative number.
If you land on a blank square stay on that square until your next turn.
The first player to reach 100 is the winner.

When You Have Finished

Find out what happens when you multiply and divide negative numbers.
Use the calculator.

100	99 (-6)	98	97 (-10)	96	95 (-5)	94	93 (-11)	92	91
81	82 (-10)	83	84 (-6)	85	86 (-11)	87	88 (-5)	89	90 (-12)
80	79	78 (-4)	77	76 (-11)	75	74	73	72 (-12)	71
61 (-3)	62	63 (-10)	64	65	66 (-4)	67 (-11)	68	69	70 (-3)
60	59 (-9)	58	57 (-4)	56	55 (-11)	54	53	52 (-8)	51
41 (-3)	42	43 (-9)	44	45 (-4)	46	47 (-6)	48 (-5)	49	50
40	39 (-8)	38	37 (-6)	36	35 (-5)	34	33	32 (-7)	31 (-5)
21 (-2)	22	23 (-10)	24 (-4)	25	26	27 (-8)	28	29 (-3)	30
20	19 (-9)	18	17	16 (-8)	15 (-1)	14	13 (-2)	12 (-8)	11
1	2 (-2)	3	4 (-3)	5	6	7 (-6)	8 (-6)	9	10 (-7)

Power of Ten

You Need

1 die
1 counter for each player
A calculator

Rules

Take turns rolling the die and move your counter. If you land on a square with a problem, work it out mentally and check your answer with a calculator.
If you were correct, roll the die again. If you were incorrect, stop there. If you land on a blank square stay on that square until your next turn.
The first player to reach 100 is the winner.

When You Have Finished

Can you think of a rule for multiplying and dividing numbers which are powers of 10?

100	99 8,000 x 800	98	97 10,000 ÷ 100	96	95 6,000 x 300	94	93	92 5,000 x 500	91
81 2,000 x 800	82	83 6,000 ÷ 3000	84	85 1,000 ÷ 200	86	87	88 5,000 x 200	89	90
80	79 5,000 ÷ 1,000	78	77 1,000 x 1,000	76	75	74 3,000 x 100	73	72	71 6,000 ÷ 2000
61 2,000 ÷ 400	62	63 3,000 x 50	64	65	66 3,000 x 50	67	68	69 5,000 ÷ 1,000	70
60	59 1,000 x 10	58	57	56 4,000 x 10	55	54	53 5,000 ÷ 200	52	51 6,000 ÷ 200
41 700 x 50	42	43	44 800 x 30	45	46	47 1,000 ÷ 10	48	49 2,000 ÷ 100	50
40	39	38 500 x 50	37	36	35 1,000 ÷ 100	34	33 800 ÷ 20	32	31
21	22 300 x 20	23	24	25 200 ÷ 50	26	27 300 ÷ 30	28	29	30 100 x 20
20 600 x 50	19	18	17 300 ÷ 50	16	15 600 ÷ 30	14	13	12 50 x 30	11
1	2	3 10x50	4	5 10 x 80	6	7	8 100 ÷ 10	9	10 100 x 50

Percentage Chase

You Need

1 die
1 counter for each player
A calculator

Rules

Take turns to roll the die and move your counter. If you land on a number circle:

1. Check the number on the die and let:
   1 = 10%;
   2 = 20%;
   3 = 30%;
   4 = 40%;
   5 = 50%; and
   6 = 60%.
2. Find that percentage of the number in the circle. For example, if you throw a 5 and land on a circle with a 30 in it then you need to find 50% of 30.
3. Move forward that number of squares. (i.e., 50% of 30 = 15, move forward 15 squares).

If you land on a blank square stay on that square until your next turn. The first player to reach 100 is the winner.

When You Have Finished

What is the furthest you could move in one turn, while playing this game?

100	99	98	97	96	95	94 (10)	93	92 (10)	91
81	82 (10)	83	84 (10)	85	86 (10)	87	88 (10)	89	90 (10)
80 (10)	79	78 (20)	77	76 (10)	75	74 (20)	73	72 (10)	71
61	62 (20)	63	64 (30)	65	66 (10)	67	68 (20)	69	70 (30)
60 (10)	59	58 (30)	57	56 (20)	55	54 (10)	53	52 (30)	51
41	42 (10)	43	44 (20)	45	46 (30)	47	48 (10)	49	50 (20)
40 (30)	39	38 (20)	37	36 (10)	35	34 (30)	33	32 (20)	31
21	22 (30)	23	24 (10)	25	26 (20)	27	28 (30)	29	30 (10)
20 (20)	19	18 (10)	17	16 (30)	15	14 (20)	13	12 (10)	11
1	2 (10)	3	4 (20)	5	6 (30)	7	8 (20)	9	10 (30)

Primes and Squares

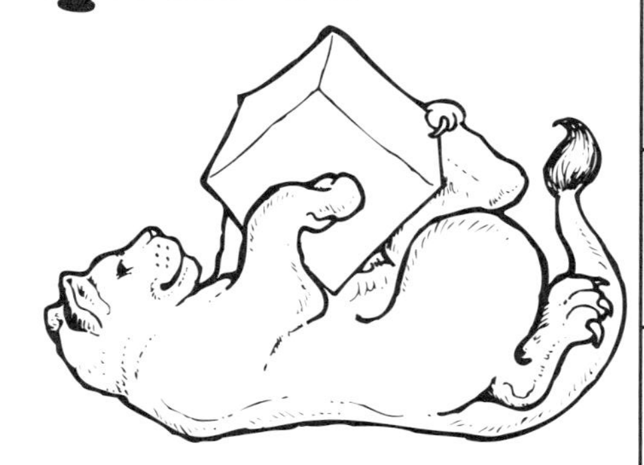

You Need
1 die
1 counter for each player

Rules
Take turns to roll the die and move your counter. If you land on one of these squares:

O - go to the next *odd number;*
E - go to the next *even number;*
P - go to the next *prime number;*
S - go to the *square* of the number; for example, land on 8(**S**) go to 64.
SR - go back to the *square root* of that number, e.g., land on 64 (**SR**) go back to 8.

Stop on that square, even if it shows another letter. If you land on a blank square stay on that square until your next turn.The first player to reach 100 is the winner.

When You Have Finished
Which numbers are the cube of 2, 3 and 4?
What is a cube root?

100 SR	99	98	97 E	96	95 P	94 E	93	92 E	91
81 SR	82	83 O	84	85 E	86	87 P	88	89 O	90 P
80 E	79 P	78	77 E	76	75 P	74 E	73	72 O	71
61 E	62	63 P	64 O	65	66 E	67	68	69 P	70 O
60	59 E	58 E	57	56 O	55 P	54	53	52 E	51
41	42 O	43	44 P	45	46 E	47	48	49 SR	50 P
40 P	39	38	37 O	36 E	35	34 P	33	32 O	31
21 E	22	23	24 P	25 SR	26	27	28 P	29	30 P
20	19 E	18	17	16 O	15 P	14	13 O	12	11 E
1	2 S	3	4 S	5	6 S	7	8 S	9 SR	10

Rounders

You Need

1 die
1 counter for each player
A calculator

Rules

Take turns to roll the die and move your counter. When you land on a square, divide the number on the square by the number on the die. Use a calculator. If the answer is a whole number, stay where you are. If the answer has a remainder you must round it to the nearest whole number. If you round up, then move your counter to the square above and roll again. If you round down, move your counter to the square below and roll again. (If there is insufficient room to move your counter up or down, just roll again.)

The first player to reach 100 is the winner.

When You Have Finished

Divide 1 by 3. If you multiply your answer by 3 what should it come to?
Try it. Why does this happen?
Does it happen with any other numbers?

100	99	98	97	96	95	94	93	92	91
81	82	83	84	85	86	87	88	89	90
80	79	78	77	76	75	74	73	72	71
61	62	63	64	65	66	67	68	69	70
60	59	58	57	56	55	54	53	52	51
41	42	43	44	45	46	47	48	49	50
40	39	38	37	36	35	34	33	32	31
21	22	23	24	25	26	27	28	29	30
20	19	18	17	16	15	14	13	12	11
1	2	3	4	5	6	7	8	9	10

Where s the Ball?

▶ **You Need**

1 die
1 counter for each player

▶ **Rules**

Roll the die and move forward that number of squares. When you land on a *line and ball square* you must state where the ball is using these words…

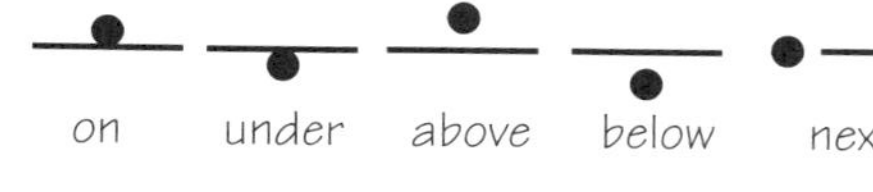

Keep this covered and use it to check if you are correct. If you are correct, roll again. If you land on a blank square, stay on that square until your next turn.
The first player to reach 36 is the winner.

▶ **When You Have Finished**

Do you know any other directions? Can you draw a line and ball to show them?

36	35	34	33	32	31
25	26	27	28	29	30
24	23	22	21	20	19
13	14	15	16	17	18
12	11	10	9	8	7
1	2	3	4	5	6

Comparisons

You Need

1 die
1 counter for each player

Rules

Take turns to roll the die and move your counter. When you land on a square containing an object, compare the object to yourself! For squares 1-40, if the object is *heavier* than you, move forward the number shown on the die. If the object is *lighter* than you, move back the number on the die. Unless you land on another object, that is the end of your turn. If you land on a blank square, stay on that square until your next turn.

For squares 41-70 the rules are the same but compare the *length* of the object with your *height*.

For squares 71-100 compare how much you can *drink* with how much water the object can *hold*.

The first player to reach 100 is the winner.

When You Have Finished

What are the heaviest and longest/tallest objects on the board? Which one has the greatest capacity (holds the most water)?

100	99 A milk bottle	98	97 A can of soda	96	95	94 A teapot	93	92	91 A bathroom sink
81	82	83 A kitchen sink	84	85 A swimming pool	86	87	88 A tea cup	89	90
80	79 A bathtub	78	77 A bucket	76	75	74 A car gas tank	73	72 A small cup	71 **Can you drink more or less than these can hold?**
61	62	63 A door	64	65	66 Half a door	67	68	69 A small bookcase	70
60	59 The school janitor	58	57 The President	56	55	54 The head teacher	53	52 A dog	51
41 **Are you taller or shorter than…**	42	43 Your mother	44	45 Your best friend	46	47	48 Your classmate	49	50 A crocodile
40	39 The head teacher	38	37 A football	36	35	34 A television	33	32 Two school chairs	31
21	22	23 Two pens	24	25 A 25¢ coin	26	27	28 Your classmate	29	30 A horse
20	19 Your best friend	18	17 Your mother	16	15	14 Your teacher	13	12 A bus	11
1 **Are you heavier or lighter than…**	2	3 A house	4	5 A pencil	6	7 A car	8 A puppy	9	10 An airplane

Shape Race

You Need
1 die
1 counter for each player

Before you start
Look at these shapes.
Can you remember their names?

square
rectangle
triangle
pentagon
hexagon

Rules
Cover the shapes with a piece of paper or fold the page back.
Take turns to roll the die and move your counter. When you land on a shape, say its name. Check your answer with the list.
If you are correct, roll again; if you are wrong, your turn is over. If you land on a blank square, stay on that square until your next turn.
The first player to reach 100 is the winner.

When You Have Finished
Talk about these questions:
How many sides does each shape have?
Can you name any other shapes?

100	99	98	97	96	95	94	93	92	91
81	82	83	84	85	86	87	88	89	90
80	79	78	77	76	75	74	73	72	71
61	62	63	64	65	66	67	68	69	70
60	59	58	57	56	55	54	53	52	51
41	42	43	44	45	46	47	48	49	50
40	39	38	37	36	35	34	33	32	31
21	22	23	24	25	26	27	28	29	30
20	19	18	17	16	15	14	13	12	11
1	2	3	4	5	6	7	8	9	10

Name that Shape (3-D)

▶ **You Need**

1 die

1 counter for each player

▶ **Before you start**

Look at these shapes.

cube

cuboid

cylinder

cone

sphere

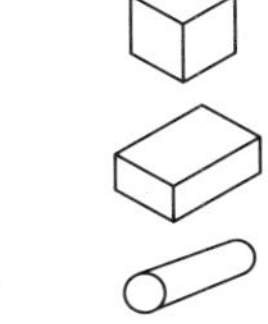

▶ **Rules**

Cover the shapes with a piece of paper or fold the page back.

Take turns to roll the die and move your counter. When you land on an object or word, say what shape it is.

Check your answer with the list. If you are correct, roll again; if you are wrong your turn is over. If you land on a blank square, stay on that square until your next turn.

The first player to reach 100 is the winner.

▶ **When You Have Finished**

Talk about these questions:

How many faces (surfaces) does each shape have?

How many edges (sides) does each shape have?

100	99	98 marble	97	96	95	94	93 Rubik's cube	92	91
81	82	83	84	85	86	87	88	89	90
80	79	78	77 book	76	75	74	73	72	71
61	62	63	64	65 tube	66 ice-cream cone	67	68	69	70
60	59	58	57	56	55	54	53	52	51
41	42	43	44 cereal box	45	46	47	48	49	50
40	39	38	37	36	35	34	33 witch's hat	32	31
21	22	23	24	25	26	27 marble	28	29	30
20	19	18	17	16	15	14	13	12 a die	11
1	2 book	3	4	5 football	6	7	8	9	10

Count the Points

You Need
1 die
1 counter for each player

Rules
Take turns to roll the die and move your counter. When you land on a shape, count the number of vertices (points) it has. Vertices are points on a shape where lines meet. For example:

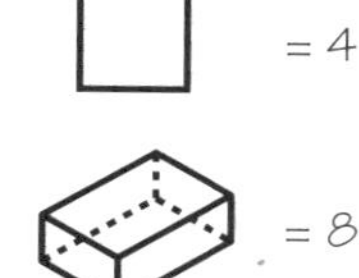

= 4 vertices

= 8 vertices

Move forward that number. Keep rolling the die until you land on a blank square. If you land on a blank square, stay on that square until your next turn. The first player to reach 100 is the winner.

When You Have Finished
Choose four different shapes and count the number of faces, edges and vertices on each. Is there any link between them?

100	99	98	97	96	95	94	93	92	91
81	82	83	84	85	86	87	88	89	90
80	79	78	77	76	75	74	73	72	71
61	62	63	64	65	66	67	68	69	70
60	59	58	57	56	55	54	53	52	51
41	42	43	44	45	46	47	48	49	50
40	39	38	37	36	35	34	33	32	31
21	22	23	24	25	26	27	28	29	30
20	19	18	17	16	15	14	13	12	11
1	2	3	4	5	6	7	8	9	10

Count the Faces

▶ **You Need**

1 die
1 counter for each player

▶ **Rules**

Take turns to roll the die and move your counter. When you land on a shape, count the number of faces and move forward that number.

For example, a [cuboid] has six faces, so move forward six squares. Keep rolling the die until you land on a blank square. If you land on a blank square, stay on that square until your next turn.
The first player to reach 100 is the winner.

▶ **When You Have Finished**

How many edges does each shape have? Can you name any of these shapes? Here are some words to help you:

prism, pyramid, triangular, cylinder, cone, cuboid, cube.

100	99	98	97	96	95	94	93	92	91
81	82	83	84	85	86	87	88	89	90
80	79	78	77	76	75	74	73	72	71
61	62	63	64	65	66	67	68	69	70
60	59	58	57	56	55	54	53	52	51
41	42	43	44	45	46	47	48	49	50
40	39	38	37	36	35	34	33	32	31
21	22	23	24	25	26	27	28	29	30
20	19	18	17	16	15	14	13	12	11
1	2	3	4	5	6	7	8	9	10

Count the Edges

100	99	98	97	96	95	94	93	92	91
81	82	83	84	85	86	87	88	89	90
80	79	78	77	76	75	74	73	72	71
61	62	63	64	65	66	67	68	69	70
60	59	58	57	56	55	54	53	52	51
41	42	43	44	45	46	47	48	49	50
40	39	38	37	36	35	34	33	32	31
21	22	23	24	25	26	27	28	29	30
20	19	18	17	16	15	14	13	12	11
1	2	3	4	5	6	7	8	9	10

You Need

1 die for each player
1 counter for each player

Rules

Take turns to roll the die and move your counter. When you land on a shape, count the number of edges it has and move forward that number. For example, a □ has four edges, a [pyramid] has eight edges. Keep rolling the die until you land on a blank square. If you land on a blank square, stay on that square until your next turn. The first player to reach 100 is the winner.

When You Have Finished

Can you name all the shapes in this game?
Can you make up names for these shapes:

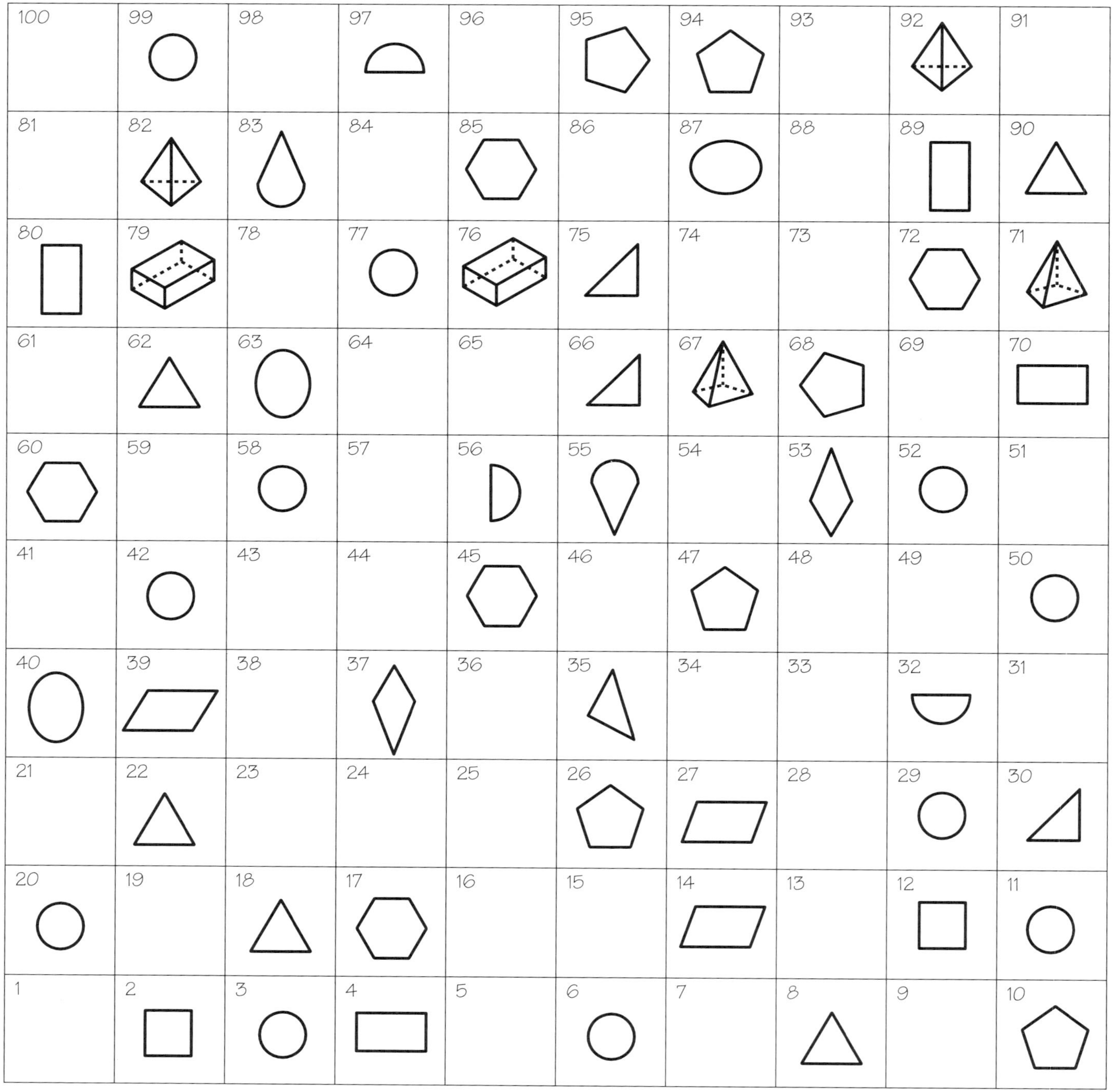

Take Turns

You Need

1 counter with an arrow on it

1 die

Rules

Before each turn, check that the arrow on your counter is pointing up. Rcll the die and move forward that number.

When you land on a 'Take Turns' square, turn your counter either 1, 2 or 3 quarter turns like this:

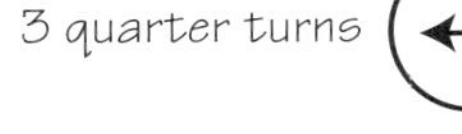

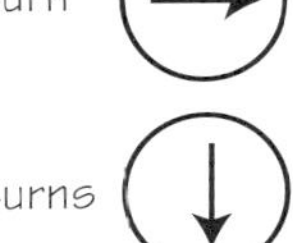

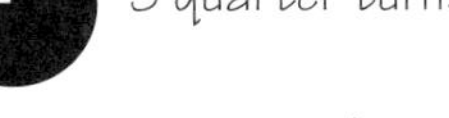

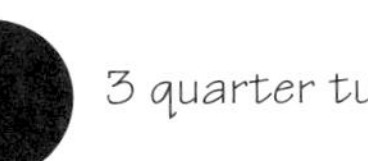

1 quarter turn

2 quarter turns

3 quarter turns

Now, move your counter the number of squares shown on the die in the direction of the arrow. If you land on a blank square, stay on that square until your next turn. The first player to reach 100 is the winner.

When You Have Finished

Can you think of another way to say 2 quarter turns? What would you call 4 quarter turns?

100	99	98	97	96	95	94	93	92	91
81	82	83	84	85	86	87	88	89	90
80	79	78	77	76	75	74	73	72	71
61	62	63	64	65	66	67	68	69	70
60	59	58	57	56	55	54	53	52	51
41	42	43	44	45	46	47	48	49	50
40	39	38	37	36	35	34	33	32	31
21	22	23	24	25	26	27	28	29	30
20	19	18	17	16	15	14	13	12	11
1	2	3	4	5	6	7	8	9	10

Right Angles

100	99	98	97	96	95	94	93	92	91
81	82	83	84	85	86	87	88	89	90
80	79	78	77	76	75	74	73	72	71
61	62	63	64	65	66	67	68	69	70
60	59	58	57	56	55	54	53	52	51
41	42	43	44	45	46	47	48	49	50
40	39	38	37	36	35	34	33	32	31
21	22	23	24	25	26	27	28	29	30
20	19	18	17	16	15	14	13	12	11
1	2	3	4	5	6	7	8	9	10

You Need

1 die
1 counter for each player

Rules

Take turns to roll the die and move your counter. When you land on a shape square, count the number of right angles in the shape and move forward that number. Then roll again. Remember! Some shapes have **no** right angles, so stay on the square and roll again. If you land on a blank square, stay on that square until your next turn. The first player to reach 100 is the winner.

When You Have Finished

What is the greatest number of right angles any four-sided shape can have? Why is this?

Symmetry

100	99	98	97	96	95	94	93	92	91
81	82	83	84	85	86	87	88	89	90
80	79	78	77	76	75	74	73	72	71
61	62	63	64	65	66	67	68	69	70
60	59	58	57	56	55	54	53	52	51
41	42	43	44	45	46	47	48	49	50
40	39	38	37	36	35	34	33	32	31
21	22	23	24	25	26	27	28	29	30
20	19	18	17	16	15	14	13	12	11
1	2	3	4	5	6	7	8	9	10

You Need

1 die
1 counter for each player

Rules

Take turns to roll the die and move your counter.
Some squares contain a number which has been reflected, as if a mirror was next to it.
For example:
the number 3 looks like this .
Sometimes the number, with its reflection, has been turned around: so you have to look carefully!
If you land on a square with a mirror number, find the number and move forward that many spaces. If you land on a blank square, stay on that square until your next turn.
The first player to reach 100 is the winner.

When You Have Finished

How many letters of the alphabet can you split into two identical parts by drawing a line through them? For example, H.

Congruent Shapes

You Need

1 die
1 counter for each player

Rules

Congruent shapes have sides the same length and the same angles. If you placed one on top of the other it would fit exactly. Take turns to roll the die and move your counter. If you land on a shape check to see if it is congruent with one of these:

If it is, roll again. Remember! Some shapes are congruent, but have been turned around.

Some shapes look similar but they are not congruent.

If you land on a blank square, stay on that square until your next turn.The first player to reach 100 is the winner.

When You Have Finished

Similar shapes have the same angles, so they look alike, but the length of the sides is different. Can you find some of these similar shapes in the game?

100	99	98	97	96	95	94	93	92	91
81	82	83	84	85	86	87	88	89	90
80	79	78	77	76	75	74	73	72	71
61	62	63	64	65	66	67	68	69	70
60	59	58	57	56	55	54	53	52	51
41	42	43	44	45	46	47	48	49	50
40	39	38	37	36	35	34	33	32	31
21	22	23	24	25	26	27	28	29	30
20	19	18	17	16	15	14	13	12	11
1	2	3	4	5	6	7	8	9	10

Rotational Symmetry

You Need

1 die
1 counter for each player

Rules

Every shape can be picked up, turned 360° and put down again in exactly the same place. This is rotating it once. Some shapes can be turned halfway and put down to make the same shape.

and again

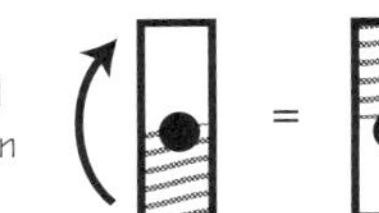

These have a rotational symmetry of two. Other shapes can have a rotational symmetry of three, four, or even more. Take turns to roll the die and move your counter. When you land on a shape count its rotational symmetry and move forward that number. Roll again. If you land on a blank square, stay on that square until your next turn.
First player to reach 100 is the winner.

When You Have Finished

Can you think of a rule for deciding the rotational symmetry of any shape? Try it out.

100	99	98 U	97	96 E	95	94 A	93 H	92	91 T
81	82	83 8	84 +	85	86 I	87	88	89 I	90
80	79	78	77 (pentagon)	76	75	74 (hexagon)	73	72	71
61	62	63	64	65 T	66	67	68	69	70
60	59	58 P	57 (hexagon)	56	55 S	54 (pentagon)	53	52	51
41	42	43	44	45	46	47	48	49	50
40 8	39	38	37 (pentagon)	36	35	34	33 (rhombus)	32	31
21	22	23	24	25	26	27	28	29	30
20	19	18	17	16	15	14	13	12	11
1	2	3	4	5	6	7	8	9	10

Choosing Instruments

You Need

1 die
1 counter for each player

Rules

Take turns to roll the die and move your counter. Look at the list of measuring instruments below, then cover them up. When you land on an item, decide which instrument you would use to measure it. For example, if you land on square 2 **weight of rice** you should decide **kitchen scales**. Check your answer on the grid below. If you are correct, roll again. If you are incorrect, that is the end of your turn.
The first player to reach 100 is the winner.

Instrument	Item Number			
Kitchen Scales	2	26	50	73
Bathroom Scales	5	29	53	76
Ruler	8	33	56	79
Meter Stick	11	36	59	83
Trundle Wheel	14	39	61	86
Stopwatch	17	41	64	89
1 liter Measuring Jug	20	44	67	94
100 ml Measuring cup	23	47	70	97

When You Have Finished

Make a list of all the other measuring instruments you know.

100	99	98	97 capacity of a spoon	96	95	94 amount of cola in a bottle	93	92	91
81	82	83 length of your leg	84	85	86 length of your house	87	88	89 length of a relay race	90
80	79 length of a matchbox	78	77	76 weight of six bricks	75	74	73 weight of a can of beans	72	71
61 length of a car parking space	62	63	64 speed of a car	65	66	67 amount of water in a bowl	68	69	70 amount of milk on a spoon
60	59 length of your arm	58	57	56 length of your hand	55	54	53 weight of your mother	52	51
41 speed of a train	42	43	44 liquid in a cup of tea	45	46	47 capacity of three drops of water	48	49	50 weight of 20 candies
40	39 length of a football pass	38	37	36 length of your desk	35	34	33 length of your finger	32	31
21	22	23 amount of water in 10 raindrops	24	25	26 weight of 10 pencils	27	28	29 weight of your teacher	30
20 amount of liquid in 20 glasses	19	18	17 length of a sprint race	16	15	14 length of the playground	13	12	11 height of a door
1	2 weight of rice	3	4	5 your own weight	6	7	8 length of a pencil	9	10

Points of the Compass

▶ **You Need**

1 die
1 counter for each player

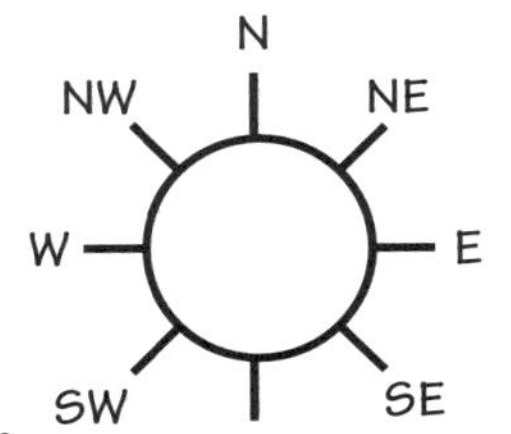

▶ **Rules**

This game uses the eight points of the compass.

Before you begin, what do these letters mean? Take turns to roll the die and move your counter. If you land on a square with a compass point, move the number of squares shown, in that direction. For example, if you land on a square which says N3, you move north three squares. If you land on a square which says SW2 you move south-west (diagonally) 2 squares. Keep rolling the die until you land on a blank square. If you land on a blank square, stay on that square until your next turn. First player to reach 100 is the winner.

▶ **When You Have Finished**

Can you name the compass points between these eight compass points?

100	99 SE1	98	97	96 W2	95 E3	94	93 S2	92 SE1	91
81 SE3	82 E2	83	84 NE1	85 SE2	86	87 SW1	88 NW1	89 NE1	90
80 N1	79	78	77 W2	76 E3	75	74 S2	73	72 SW2	71 N1
61	62 W1	63 E3	64	65 NW1	66 NE1	67	68 S2	69 SE1	70
60 N1	59	58 E4	57	56 W2	55 NW1	54	53 NE1	52 W2	51
41 E3	42 NE2	43	44 N1	45 S2	46	47 S1	48	49 N3	50 NW1
40	39 NW1	38 W2	37	36 NE3	35 S2	34	33 NW2	32 SW2	31 S3
21 NE1	22	23 NW1	24 W2	25	26 SW2	27 N1	28	29 S1	30 W5
20	19 S1	18	17 E3	16 N2	15	14	13 W4	12	11 W3
1	2 N2	3 N1	4	5 E3	6 E3	7 W2	8	9 N2	10

Lines of Symmetry

You Need

1 die
1 counter for each player

Rules

Take turns to roll the die and move your counter. If you land on a shape, decide how many lines of symmetry it has and move forward that number (the first row has the lines drawn in for you). End your turn there.
If you land on a blank square, stay on that square until your next turn.
If you disagree on the number of lines of symmetry, work the answer out together.
The first player to reach 100 is the winner.

When You Have Finished

Which four-sided shape has the most lines of symmetry?
Which three-sided shape has the most lines of symmetry.

100	99	98	97	96	95	94	93	92	91
81	82	83	84	85	86	87	88 X	89	90
80	79	78	77	76	75	74	73	72	71
61	62 X	63	64	65	66 Y	67	68	69	70
60 W	59	58 V	57	56 M	55	54	53 8	52	51 3
41	42 C	43	44 D	45	46 E	47	48	49 I	50
40	39	38 8	37	36 A	35	34	33 H	32	31 T
21	22	23	24	25	26	27	28	29	30
20	19	18	17	16	15	14	13	12	11
1	2	3	4	5	6	7	8	9	10

Name that Angle

Wow! That's a cute angle!

You Need

1 die
1 counter for each player

Rules

Take turns to roll the die and move your counter. When you land on a square which contains an angle, you have to name the angle.

Here are their names:

acute angle	less than 90°
right angle	exactly 90°
obtuse angle	greater than 90° and less than 180°
reflex angle	greater than 180°

Keep these answers covered while you play. If you name the angle correctly, take another turn. If you land on a blank square, stay on that square until your next turn. The first player to reach 100 is the winner.

When You Have Finished

What is the biggest angle you could make?
What is the smallest angle you could make?
Is 180° an angle?

100	99	98	97	96	95	94	93	92	91
81	82	83	84	85	86	87	88	89	90
80	79	78	77	76	75	74	73	72	71
61	62	63	64	65	66	67	68	69	70
60	59	58	57	56	55	54	53	52	51
41	42	43	44	45	46	47	48	49	50
40	39	38	37	36	35	34	33	32	31
21	22	23	24	25	26	27	28	29	30
20	19	18	17	16	15	14	13	12	11
1	2	3	4	5	6	7	8	9	10

Millimeters, Centimeters, Meters and Kilometers

You Need

1 *die*
1 *counter for each player*
A *calculator*

Rules

Take turns to roll the *die* and move your counter. If you land on a circle you must change the measurement inside it to *kilometers*. A square to *meters*; hexagon to *centimeters*; triangle to *millimeters*. Write down your answer, then check it on the grid below. If you are correct, roll again. If you land on a blank square, stay on that square until your next turn. Keep the answer grid covered while playing.
The first player to reach 100 is the winner.

When You Have Finished

Can you think of a rule/s to help convert these metric units?

2 km = 2,000 m	1 km = 1,000 m	0.5 km = 500 m	200 cm = 2 m
1 m = 100 cm	0.5 m = 50 cm	20 mm = 2 cm	2.5 m = 250 cm
1 cm = 10 mm	1 m = 1,000 mm	5 cm = 50 mm	0.5 cm = 5 mm
1,000 m= 1 km	2,000 m = 2 km	1,500 m = 1.5 km	500 m = 0.5 km

100	99	98 2 km	97	96 500 m	95	94 0.5 cm	93	92 2.5 m	91
81	82 0.5 km	83	84 20 mm	85	86 5 cm	87	88 1,500 m	89	90 1 km
80 1,000 m	79	78 1 m	77	76 0.5 m	75	74 200 cm	73	72 2,000 m	71
61	62 1 cm	63	64 500 m	65	66 2 km	67	68 1 m	69	70 0.5 cm
60 20 mm	59	58 1 km	57	56 1,500 m	55	54 5 cm	53	52 2.5 m	51
41	42 200 cm	43	44 0.5 m	45	46 1 m	47	48 2,000 m	49	50 0.5 km
40 100 m	39	38 1 cm	37	36 1 m	35	34 2 km	33	32 500 m	31
21	22 5 cm	23	24 1,500 m	25	26 1 km	27	28 2.5 m	29	30 0.5 cm
20 20 mm	19	18 0.5 km	17	16 2,000 m	15	14 1 m	13	12 0.5 m	11
1	2 2 km	3	4 1 m	5	6 1 cm	7	8 1,000 m	9	10 200 cm

Converting Liters and Milliliters

You Need

1 die
1 counter for each player
A calculator

Rules

Take turns to roll the die and move your counter. If you land on a liter measurement (e.g., 3 L) divide the measurement by the number on the die. Write your answer in milliliters and check it on the grid below. If you land on a milliliter measurement (e.g., 500 ml) multiply the measurement by the number on the die. Write down your answer in liters and check it on the grid below. If correct, roll the die again. If you land on a blank square, stay on that square until your next turn.

Before you start, look at the answers in the box below and try a few examples. The first player to reach 100 is the winner.

Die	3 l	6 l	9 l	500ml	800ml	30ml
1	3,000 ml	6,000 ml	9,000 ml	0.5 l	0.8 l	0.03 l
2	1,500 ml	3,000 ml	4,500 ml	1 l	1.6 l	0.06 l
3	1,000 ml	2,000 ml	3,000 ml	1.5 l	2.4 l	0.09 l
4	750 ml	1,500 ml	2,250 ml	2 l	3.2 l	0.12 l
5	600 ml	1,200 ml	1,800 ml	2.5 l	4 l	0.15 l
6	500 ml	1,000 ml	1,500 ml	3 l	4.8 l	0.18 l

When You Have Finished

Are 0.5 l, 0.50 l, 0.005 l and 0.05 l the same? Write each measurement in milliliters.

100	99 **800 ml**	98	97	96 **9 L**	95	94	93 **500 ml**	92	91
81	82	83 **3 L**	84	85	86 **30 ml**	87	88	89 **6 L**	90
80	79 **800 ml**	78	77	76 **9 L**	75	74	73 **500 ml**	72	71
61	62	63 **3 L**	64	65	66 **30 ml**	67	68	69 **6 L**	70
60	59 **800 ml**	58	57	56 **14**	55	54	53 **500 ml**	52	51
41	42	43 **3 L**	44	45	46 **30 ml**	47	48	49 **6 L**	50
40	39 **800 ml**	38	37	36 **9 L**	35	34	33 **500 ml**	32	31
21	22	23 **3 L**	24	25	26 **30 ml**	27	28	29 **6 L**	30
20	19 **800 ml**	18	17	16 **9 L**	15	14	13 **500 ml**	12	11
1	2	3 **3 L**	4	5	6 **30 ml**	7	8	9 **6 L**	10

Converting Kilograms and Grams

You Need

1 die
1 counter for each player
A calculator

Rules

Take turns to roll the die and move your counter. If you land on a kilogram measurement (e.g., 3 kg) divide the measurement by the number on the die. Write your answer in grams and check it on the grid below. If you land on a gram measurement (e.g., 500 g) multiply the measurement by the number on the die. Write down your answer in kilograms and check it on the grid below. If correct, roll the die again. If you land on a blank square, stay on that square until your next turn. Before you start, look at the answers in the box below and try a few examples. The first player to reach 100 is the winner.

Die	3kg	6kg	9kg	500g	800g	30g
1	3,000 g	6,000 g	9,000 g	0.5k g	0.8 kg	0.03 kg
2	1,500 g	3,000 g	4,500 g	1k g	1.6 kg	0.06 kg
3	1,000 g	2,000 g	3,000 g	1.5 kg	2.4 kg	0.09 kg
4	750 g	1,500 g	2,250 g	2 kg	3.2kg	0.12 kg
5	600 g	1,200 g	1,800 g	2.5 kg	4 kg	0.15 kg
6	500 g	1,000 g	1,500 g	3 kg	4.8 kg	0.18 kg

When You Have Finished

Are 0.5 kg, 0.50 kg, 0.005 kg and 0.05 kg the same? Write each measurement in grams.

100	99 800 g	98	97	96 9 kg	95	94	93 500 g	92	91
81	82	83 3 kg	84	85	86 30 g	87	88	89 6 kg	90
80	79 800 g	78	77	76 9 kg	75	74	73 500 g	72	71
61	62	63 3 kg	64	65	66 30 g	67	68	69 6 kg	70
60	59 800 g	58	57	56 9 kg	55	54	53 500 g	52	51
41	42	43 3 kg	44	45	46 30 g	47	48	49 6 kg	50
40	39 800 g	38	37	36 9 kg	35	34	33 500 g	32	31
21	22	23 3 kg	24	25	26 30 g	27	28	29 6 kg	30
20	19 800 g	18	17	16 9 kg	15	14	13 500 g	12	11
1	2	3 3 kg	4	5	6 30 g	7	8	9 6 kg	10

Imperial and Metric

You Need

1 die
1 counter for each player

Rules

Look at the metric and imperial measures on the grid below. Keep this covered during the game and use it to check your answers.
Take turns to roll the die and move your counter. When you land on a metric measure, give its imperial equivalent. (Check your answer.)
When you land on an imperial measure, give its metric equivalent. (Check your answer.)
If you are correct, take another turn. If you land on a blank square, stay on that square until your next turn. If you are incorrect, end your turn.

1 kg	=	2 pounds
1 pound	=	$^1/_2$ kilogram
1 mile	=	1.6 kilometer
1 kilometer	=	$^2/_3$ mile
1 liter	=	1.75 pints
1 gallon	=	$4^1/_2$ liters

(Note: conversions are approximate.)
The first player to reach 100 is the winner.

When You Have Finished

Do you know any other imperial measurements?
When do we still use imperial measures?

100	99	98 1 kg	97	96	95 1 mile	94	93	92 1 liter	91
81	82 1 pound	83	84	85 1 km	86	87	88 1 gallon	89	90
80 2 pounds	79	78	77 $\frac{1}{2}$ kg	76	75	74 1.6 km	73	72	71 $\frac{2}{3}$ mile
61	62	63 1.75 pints	64	65	66 $4\frac{1}{2}$ liters	67	68	69 1 kg	70
60	59 1 mile	58	57	56 1 liter	55	54	53 1 pound	52	51
41 1 km	42	43	44 1 gallon	45	46	47 2 pounds	48	49	50 $\frac{1}{2}$ kg
40	39	38 1.6 km	37	36	35 $\frac{2}{3}$ mile	34	33	32 1.75 pints	31
21	22 $4\frac{1}{2}$ liters	23	24	25 1 kg	26	27	28 1 mile	29	30
20 1 liter	19	18	17 1 pound	16	15	14 1 km	13	12	11 1 gallon
1	2	3 2 pounds	4	5	6 $\frac{1}{2}$ kg	7	8	9 1.6 km	10

Estimating

You Need

1 die
1 counter for each player

Rules

Take turns to roll the die and move your counter. When you land on a problem square, decide if the estimate is a sensible one. Check your answer with the grid below. All other answers are wrong! If you are correct, roll again. If you are incorrect, stay where you are. If you land on a blank square, stay on that square until your next turn. The first player to reach 100 is the winner.

football pass	100 m	tea cup	250 ml
car gas tank	50 L	beer glass	0.5 L
book	30 cm	boy	20 kg
bottle of cola	2 L	chocolate bar	200 g
brick	0.5 kg	marathon race	40 km
desk	10 kg	a running track	400 m
pencil	20 cm	bucket	5 L
compact disc	100 g	fly	15 mm
eraser	50 g	bathtub	100 L

When You Have Finished

Look at each of the items in the table above and decide what is a good range for acceptable estimates of each one.
For example, would 80 m-120 m be a good range for a football pass?
Or 50 m-200 m?

100	99 boy **500 g**	98	97	96 beer glass **0.5 L**	95 fly **0.5 m**	94	93 tea cup **20 mL**	92	91 chocolate bar **200 g**
81	82	83 desk **10 kg**	84	85	86 compact disc **1 kg**	87	88	89 running track **400 m**	90
80 bottle of cola **20 mL**	79	78	77 car gas tank **50L**	76	75	74 football pass **100 km**	73	72	71 bathtub **100 L**
61	62 bucket **100 mL**	63	64	65 pencil **20 cm**	66 fly **15 mm**	67	68 rubber **50 kg**	69	70 tea cup **2 L**
60 bath **10 L**	59	58 brick **0.5k g**	57	56	55 book **10 m**	54	53	52 fly **15 cm**	51
41 beer glass **10 L**	42	43	44 weight of a boy **20 kg**	45	46	47 chocolate bar **10 kg**	48	49	50 marathon race **40 km**
40	39 tea cup **250 mL**	38	37	36 running track **10 m**	35 marathon race **400 cms**	34	33 compact disc **100 g**	32	31 bucket **100 L**
21 weight of a boy **500 kg**	22	23 car gas tank **500 L**	24	25	26 bottle of cola **2 L**	27	28	29 desk **100 kg**	30
20 football pass **100 m**	19	18	17 capacity of a bathtub **5 L**	16	15	14 weight of an eraser **50 g**	13	12	11 length of a pencil **20 km**
1	2 length of a book **30 cm**	3	4	5 weight of a brick **1 g**	6 weight of 6 candies **200 g**	7	8 capacity of a bucket **30 cm**	9	10

Angles of Rotation

You Need

1 die
1 counter for each player with an arrow on it like this:

Rules

Keep the arrow pointing upwards while you move around the board. Take turns to roll the die and move your counter. When you land on an angle square, turn the arrow on your counter (clockwise) the number of degrees shown. Then move onto the next square in the direction shown by the arrow and roll the die again.

For example: if you land on [90°], turn your arrow counter through 90° (↑)-(→) and move to the next square in that direction. Then roll again. If you land on a blank square, stay on that square until your next turn. The first player to reach 100 is the winner.

When You Have Finished

What are the names of these angles?
Less than 90°
Between 91° and 189°
Between 181° and 360°
Exactly 90°

100	99 135°	98	97 270°	96	95 135°	94	93 135°	92	91 225°
81 90°	82 45°	83	84 180°	85	86 270°	87	88 360°	89	90 180°
80	79	78 225°	77	76 225°	75	74 225°	73	72 225°	71
61 45°	62	63 315°	64	65 315°	66	67 315°	68	69 315°	70
60 180°	59 360°	58	57 135°	56	55 135°	54	53 135°	52	51 225°
41	42 45°	43	44 450°	45	46 45°	47	48 45°	49	50 270°
40 360°	39	38 180°	37	36 270°	35	34 90°	33	32 360°	31
21	22 90°	23	24 90°	25	26 360°	27	28 360°	29	30 270°
20 360°	19	18 270°	17	16 180°	15	14 270°	13	12 360°	11
1	2 90°	3	4 360°	5	6 90°	7	8 360°	9	10 360°

Name those Lines

You Need

1 die
1 counter for each player

Rules

Take turns to roll the die and move your counter. When you land on a square with a line or lines on it, you must say what they are. You have a choice of 6 answers.

Horizontal	**Vertical**
Parallel	**Perpendicular**
Intersecting	**Converging**

Keep this box covered during the game until you need to check your answer.

If you are correct, roll again. If you are incorrect, end your turn. If you land on a blank square, stay on that square until your next turn.

The first player to reach 100 is the winner.

When You Have Finished

Make a list of where you can find each of these lines in the classroom.

100	99	98	97	96	95	94	93	92	91
81	82	83	84	85	86	87	88	89	90
80	79	78	77	76	75	74	73	72	71
61	62	63	64	65	66	67	68	69	70
60	59	58	57	56	55	54	53	52	51
41	42	43	44	45	46	47	48	49	50
40	39	38	37	36	35	34	33	32	31
21	22	23	24	25	26	27	28	29	30
20	19	18	17	16	15	14	13	12	11
1	2	3	4	5	6	7	8	9	10

Quadrilaterals

You Need

1 die
1 counter for each player

Rules

Below are the names of the quadrilaterals used in this game. Make sure you know what each one looks like.

- square
- rectangle
- parallelogram
- rhombus
- kite
- trapezium
- Isosceles trapezium

Take turns to roll the die and move your counter. When you land on a square with a quadrilateral, you must name it. If your opponent challenges what you have said then you must discuss who is correct. If you are correct, take another turn. If you are wrong, that is the end of your turn. If you land on a blank square, stay on that square until your next turn.
The first player to reach 100 is the winner.

When You Have Finished

Which of these shapes are related? They all have 4 sides but some of them are also related in other ways - the length of their sides, for example. Can you find them?

100	99	98	97	96	95	94	93	92	91
81	82	83	84	85	86	87	88	89	90
80	79	78	77	76	75	74	73	72	71
61	62	63	64	65	66	67	68	69	70
60	59	58	57	56	55	54	53	52	51
41	42	43	44	45	46	47	48	49	50
40	39	38	37	36	35	34	33	32	31
21	22	23	24	25	26	27	28	29	30
20	19	18	17	16	15	14	13	12	11
1	2	3	4	5	6	7	8	9	10